The UMAP Expository Monograph Series

UMAP Monographs bring the undergraduate student new mathematics and fresh applications of mathematics with no delay between the development of an idea and its implementation in the undergraduate curriculum.

The High Cost of Clean Water:
Models for Water Quality Management
Edward Beltrami, *SUNY at Stony Brook*

Spatial Models of Election Competition
Steven J. Brams, *New York University*

Elements of the Theory of Generalized Inverses for Matrices
Randall E. Cline, *University of Tennessee*

Introduction to Population Modeling
James C. Frauenthal, *SUNY at Stony Brook*

Smallpox: When Should Routine Vaccination be Discontinued?
James C. Frauenthal, *SUNY at Stony Brook*

Conditional Independence in Applied Probability
Paul E. Pfeiffer, *Rice University*

Topics in the Theory of Voting
Philip D. Straffin, Jr., *Beloit College*

Man in Competition with the Spruce Budworm:
An Application of Differential Equations
Philip M. Tuchinsky, *Ford Motor Company's Research and Engineering Center*

The **UMAP** Expository Monograph Series

Edward Beltrami

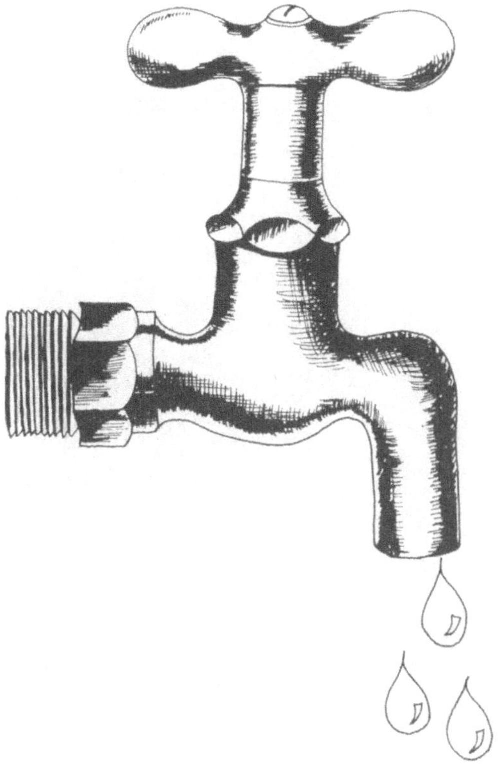

The High Cost of Clean Water

Models for Water Quality Management

Springer Science+Business Media, LLC

Author:

Edward Beltrami
Applied Mathematics and Statistics
State University of New York
Stony Brook, New York 11794

ISBN 978-0-8176-3098-0 ISBN 978-1-4899-5322-3 (eBook)
DOI 10.1007/978-1-4899-5322-3

This material was initially developed in the Department of Applied Mathematics and Statistics at State University of New York, Stony Brook (NSF Grant No. SED76-05433), and has continued development under UMAP with the partial support of National Science Foundation Grant No. SED80-07731. Recommendations expressed are those of the author and do not necessarily reflect the views of the NSF or the copyright holder.

Originally published by Education Development Center, Inc. in 1982

Table of Contents

L.I. WATER PLANNERS URGE GROWTH LIMITS

Keeping Two-Acre Zoning in Parts Held an Alternative to Sewers in Protecting Drinking Supply

By FRANCES CERRA
Special to The New York Times

HAUPPAUGE, L.I., April 28 — The cheapest way to preserve a pure supply of drinking water on Long Island is to limit the development of Nassau County's North Shore and the still-rural areas of Suffolk County, the director of the Nassau-Suffolk Regional Planning Board said today.

The director, Lee F. Koppelman, made the statement as part of the formal presentation of the findings of a two-year, $5.2 million comprehensive study of Long Island's water supply and waste-disposal problems. Because all of Long Island's drinking water comes from wells, management of that critical resource has wide implications for future development.

The study found that the installation of expensive sower systems could be avoided by other measures, primarily the preservation of two-acre zoning on some of Nassau's North Shore and in the eastern part of the town of Brookhaven in Suffolk.

The study also recommended a series of improvements in existing sewage-treatment facilities and the installation of new sewers at a total cost of $1 million.

Sewer Criterion Is Criticized

Among the measures recommended, was one that called for the relocation of the outfall from Nassau's Bay Park sewage treatment plant to the ocean, at an estimated cost of $52 million. Expanded or new sewer systems were recommended for parts of Huntington, Kings Park, the South Fork and the southern part of Suffolk County bordering the eastern portion of the Great South Bay, among other areas.

One of the major criteria for recommending new sewers, according to the study, is the presence of more than three dwelling units an acre. This criterion was criticized by the environmental consultant to the study as being lower than necessary and as arrived at by a political process.

Zoning on the North Shore of Nassau and in the eastern part of Brookhaven Town is inextricably linked to the need for sewers, according to the study.

Unlike the South Shore, where the wells are protected by impervious layers of clay from ground pollutants, a denser North Shore population without sewering could threaten that area's wells.

An Expensive Undertaking

Mr. Koppelman said that if the recommendation on two-acre zoning "is not followed and we have to have county-wide sewering, the only people who will be able to afford to live out here are the wealthy."

As shown by the example of the Southwest Sewer District, now under construction in portions of Islip and Babylon Towns, sewers are extremely expensive. The cost of that controversial system has risen from the $291 million originally estimated to $832 million, including interest.

Mr. Koppelman acknowledged that low density zoning was obviously incompatible with the housing needs of low- and moderate-income families, but he said that provision for such housing could be made in areas where sewers already existed or were under construction, such as the Southwest Sewer District.

Mr. Koppelman also said that Nassau County should be able to meet its drinking-water needs without having to resort to piping in water from off the Island.

'Politics' Are Implicated

After Mr. Koppelman's presentaion, the environmental consultant to the study told reporters that he was upset that politics "had played a heavy part" in the conclusions of the study, and had biased them toward the construction of future sewers. Specifically, he criticized the study conclusion that called for the consideration of sewers wherever there were more than three dwelling units an acre.

The consultant, Ronald Beck of the Energy Resources Company of Cambridge, Mass., said: "Five dwelling units per acre would have been a proper criterion from an environmental standpoint. The problem was that many of the voting members of the study group work for agencies that are building sewers. At the end, these agency representatives were dominated by what their superiors wanted them to do."

Mr. Beck declined to specify which agency heads had influenced their representatives.

1 Introduction

A severe problem for many regions of the nation is that of waste-water disposal. With the increase of population in urban and surburban areas and the proliferation of industrial activity, the water drawn from aquifers and reservoirs to meet these needs has also increased. After the water is used, it is either unavailable or unfit (because of contaminants) for further use. The wastes are generally discarded into neighboring water bodies (streams, bays) through sewer outfalls or discharged back into the ground (cesspools, sumps). In addition, toxic residues which accumulate on the paved surfaces of developed areas and pesticides and fertilizers on cultivated acreage, also flow into nearby water bodies as runoff after storms. The effluents which reach either the ground or surface waters may not only endanger health but also could pose a significant nuisance in terms of eutrophication. This is a condition which is characterized by an excessive growth of algae, an accumulation of organic debris, as well as low oxygen levels in the waters.

In order to cope with this problem, the Federal Water Pollution Control Act of 1972, as administered by EPA, has required (under section 208 of the Act) that a number of planning agencies throughout the country prepare regional wastewater management plans which address water quality in each of their designated areas. The guidelines to be followed by these local agencies in the preparation of a '208' management plan specify that the study should include "municipal and industrial point discharges, pollution emanating from diffuse sources, protection of groundwater, and pollution resulting from disposal of residual wastes." The plan is to be "cost-effective...and implementable." It further specifies that wastewater systems for each area be defined in terms of "system configurations, location and capacity of all facilities, treatment level and types," etc. (ref. 1).

One region particularly affected by these requirements is the Long Island area of New York (Nassau and Suffolk counties). It consists of a narrow strip of land surrounded by a large number of bays and inlets which support a substantial shellfish industry as well as extensive recreational activities. The entire water supply of the region, for a population of nearly three million, is pumped from underground glacial aquifiers and after use is either recharged back into the ground or is conveyed by sewers for disposal into the bays or the ocean beyond. Either alternative is not wholly satisfactory. Even if the effluents are treated to remove contaminants some of the pollutants will remain and, in the case of the return to groundwater, there is at least some risk of degrading the public water supply. On the other hand, discharge into the bays removes water from the aquifiers and, by not returning it, the water table is lowered. This affects stream flow (since these are largely fed by underground reservoirs) and may even induce salt water encroachment. In addition, the shellfish activities require unpolluted bay waters. A last alternative of ocean outfall still does not replenish the groundwaters, is generally quite expensive, and not often reliable. Moreover, ocean dumping may, depending on ocean currents, affect beach areas adversely.

There are essentially two ways of deflecting the impact of disposal. The first consists of land use changes (through zoning, local ordinances, and so forth) which will reduce or, at least more favorably redirect, the generation and flow of wastewaters. This is particularly significant in terms of reducing non-point sources of pollution such as runoff from developed areas. The second approach centers around the disposal of effluents carried as sewage. This means designing a configuration of treatment facilities and disposal sites which minimize the adverse effects on water quality and on groundwater quality. In this monograph we deal primarily with this structural approach to pollution abatement and leave as a classroom exercise the structural alternatives represented by shifts in land use through zoning (Appendix B). Nor do we consider other options such as limitations on fertilizer use or the banning of certain pesticides and detergents, important as these may be. Our discussion is based on the work done in the Long Island '208' study, in connection with the Nassau-Suffolk Regional Planning Board, but of course similar considerations are applicable in other coastal areas.

In the next section, the problem setting is examined in more detail with a discussion of some of the background data, including cost figures. Cost is only one of the main factors which intervene in the final acceptance of a plan, but it is perhaps the most persuasive. Since each structural alternative must eventually be judged in terms of how it affects the receiving water body we need to be able to quantify water quality. A simple approach to water pollution is outlined in Section 3 by means of a differential equation model and is then applied in the subsequent Section 4. Here a model is formulated which allows one to articulate the tradeoffs between cost and water quality in terms of various options on sewage plant locations and the diversion of flow from one service area to another. The model could also consider the different types of waste treatment which are possible for the effluents discharged by the treatment facilities, but this will not be done here.

The siting model is a mixed integer program with nonlinear objective known as a "fixed-charge plant location model." Computation with such programs can be a formidable task but, as we will see, certain intuitively plausible heuristic arguments reduce their complexity to a more tractable form. A numerical example is included for a prototype of the actual Long Island situation. Input data are flows and loads of wastewater pollutants from sources within the region. This is followed by a brief discussion of an alternate integer programming model of the same problem using what is known as a "set covering" formulation. The advantages and disadvantages of each are then compared.

In the final section, the model is discussed in terms of the problems of implementation which are encountered in regional planning. We will see that the model serves to screen some of the choices which are available to planners in the face of fiscal constraints, in terms of minimizing the threat of degradation to surface waters from the aggregate impact of land use activities scattered over a large area.

This monograph introduces the student to an important problem of national concern and it demonstrates the role of mathematical modelling in the public interest. However, in order to score certain pedagogical points we have deliberately distorted the original problem setting. The "prototype" situation discussed by us is a composite of several waste treatment options from different portions of the Long Island region. Therefore, the geographic configuration dealt with here only vaguely resembles the real thing although both the background data and the implication of model interpretation are factually quite close to those employed in the '208' study itself. A comprehensive summary of the actual Long Island '208' study is provided by Ref. 2.

There are three appendices. The first is an extended classroom exercise dealing with waste water pollution in a river while the second is also an extended exercise which introduces the possibility of zoning as a means of combating pollution in coastal waters. They are designed to give the student an opportunity to review and enlarge on the modelling concepts introduced earlier in the monograph. The third appendix provides solutions to the various exercises posed throughout the text.

2 Problem Background

Let us begin by exhibiting the actual configuration of waste treatment facilities in Nassau County (which is adjacent to the New York City line). This area is the most populated of the two counties in Long Island as well as the most developed. In Figure 1 we see that a number of facilities are already in place (as of 1976/77) and the planning options are largely a question of expanding and/or upgrading the existing plants with perhaps new hookups between a few of these to promote economy of scale. In addition, some plants may have their outfalls diverted to new locations either for purposes of recharge to ground or to where sewage effluent is less likely to contravene water quality standards.

As we move east of Nassau County the population thins out considerably and much of the land is still relatively undeveloped. Also most of the area is presently unsewered so it is possible to articulate a number of new disposal options. By dividing the county into a number of service districts (each determined by geographic considerations as well as by township boundaries) we can focus attention on five adjacent areas, some of which are proposed as sewer districts (Fig. 2). These would service a region whose south shore borders a bay with an ocean inlet (as shown) and with the north shore facing Long Island Sound. Environmental (neé sanitary) engineering consultants on the '208' project have determined, in conjunction with the planning board and local water authority and public works

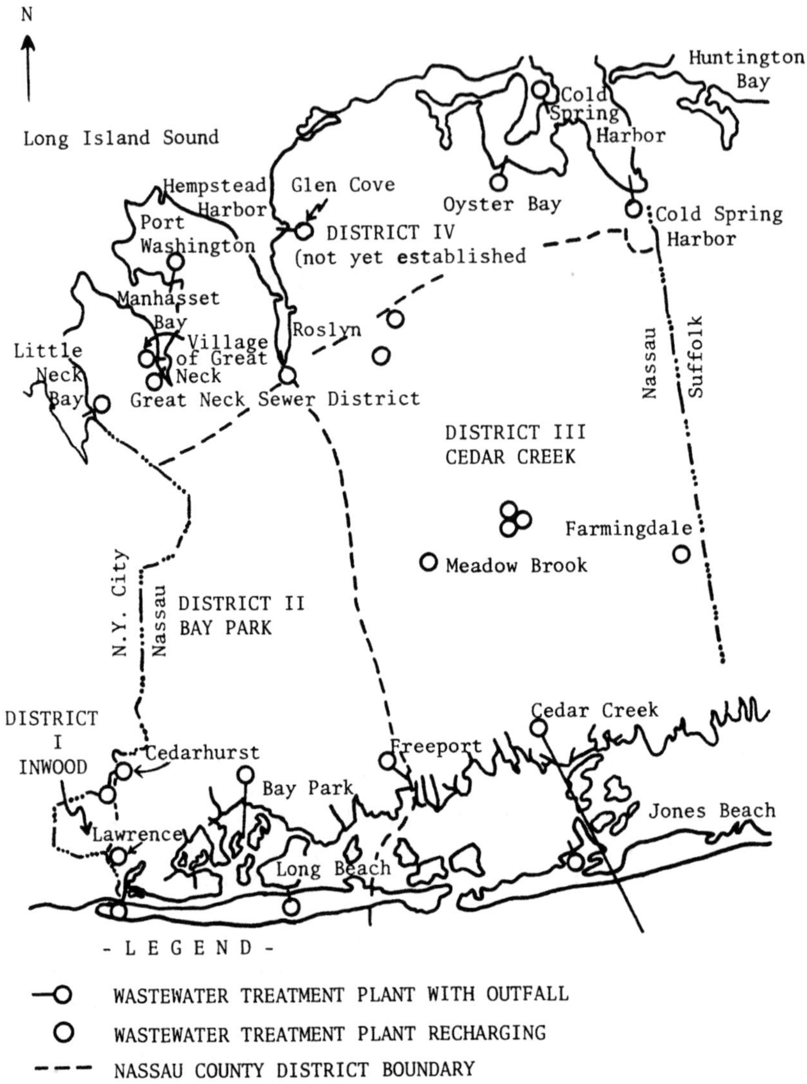

Figure 1. Nassau County Sewer Districts and Existing Wastewater Treatment Plants.

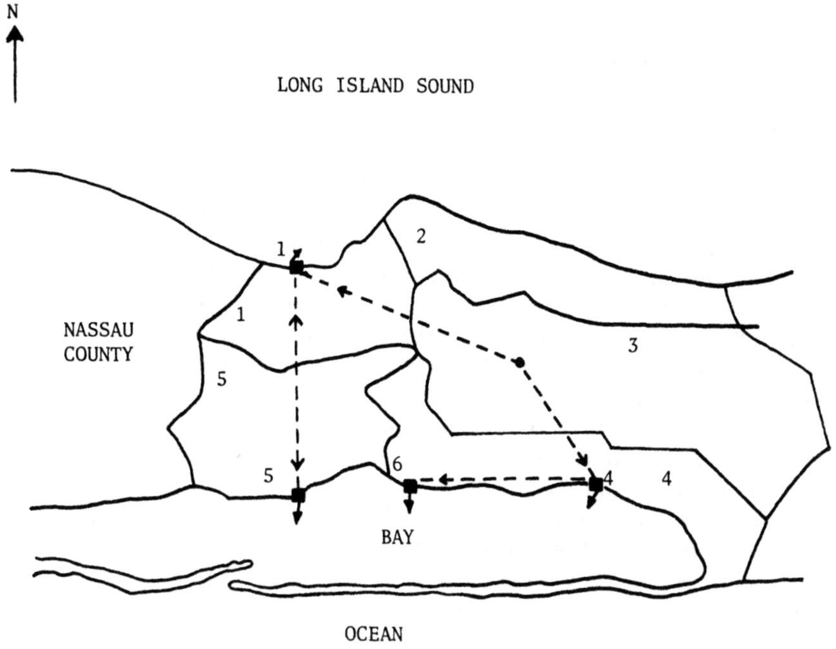

Figure 2. Configuration of Waste Treatment Facilities and Connecting Sewer Trunk Lines in Eastern Long Island.

agencies, several possible locations for new treatment plants and their outfalls, as well as the interconnections between them. These are also shown in Figure 2. We will postpone until Section 4 a detailed discussion of the specific layout and limit ourselves to some general background information here. First, it is important to reaffirm that the exhibited configuration of plants is only a prototype situation in which some of the options available to the Planning Board are presented together within a single region.

Households which are connected to sewers have their wastes fed into larger mains which are joined to the treatment plants. This "feeder" system has a construction cost which is some fixed amount, depending on the size of the sewered community. For the purposes of our study, what is more significant is the cost of the main pipelines (the "interceptors" or "trunk" sewers) which convey the wastes from the communities, since the cost of this will vary depending on the final choice of plant location(s). In fact it is one of the purposes of the model to be developed in Section 4 to determine these locations in such a way that total cost, including that of pipeline, be as small as possible. From our point of view, therefore, the construction of the local feeder system of sewers will be assumed as given and we concentrate our attention on the as yet unspecified trunk sewer lines.

We also assume that, in the surburban area being considered here, wastewater flow will be generated at the rate of 100 gallons per day for each person in the sewered areas. This also accounts indirectly for water used and discharged by the commerical enterprises and other support activities within the district (schools, hospitals, government offices, etc.) which serve this population. Total daily flow in each district is given in units of million gallons per day (mgd). For purposes of comparison with more familiar units, you may wish to note that 1 mgd is equivalent to roughly 1.55 cubic feet per second.

The construction cost C_c of a disposal plant is dependent on its size and on the kind of waste treatment it provides. Because of economies of scale, these costs increase with size at a decreasing rate. A useful approximation for our purposes (units of thousands of dollars per annum) is given by

$$C_c = 44.0\ Q^{0.8}$$

where Q is the designed capacity of the plant in mgd. The construction cost is amortized over a financing period of 25 years (generally supported through bond issues) at a specified interest rate and then computed on an annual basis. The cost, C_c, assumes primary treatment, which is essentially the removal of suspended solids in the wastewater by sedimentation. Secondary treatment, a more involved procedure which removes a larger fraction of the organic wastes by means of bacterial decomposition, is more costly. The annual figure in this case is given by

$$C_c = 52.0\ Q^{0.8}$$

(thousands of dollars). We consider only secondary treatment in our model.

One can similarly compute the expense of operating the plants and of constructing the sewer pipes which convey the wastes. Operating cost depends on such factors as maintenance and sludge removal and is not included in the above relations. This expense is usually borne by the user through local charges and property taxes. The cost of an interceptor system includes excavation and right of way compensations. Ideally a sewer system maintains flow through gravity but, depending on the slope of the terrain, hydraulic conditions may necessitate either deep excavations or forced flow uphill, the latter involving pumping stations. These combined sewer costs can be incorporated into a single annual figure C_s which we write as

$$C_s = 2.8\ Q^{0.60}$$

in units of thousands of dollars per mile of trunk sewer. It is perhaps useful to observe that the cost figures are concave functions which, when plotted on log-log paper, become straight lines. Thus, for example, the dollar costs per mgd of capacity are those displayed in Figure 3, which is the way such numbers are usually presented.

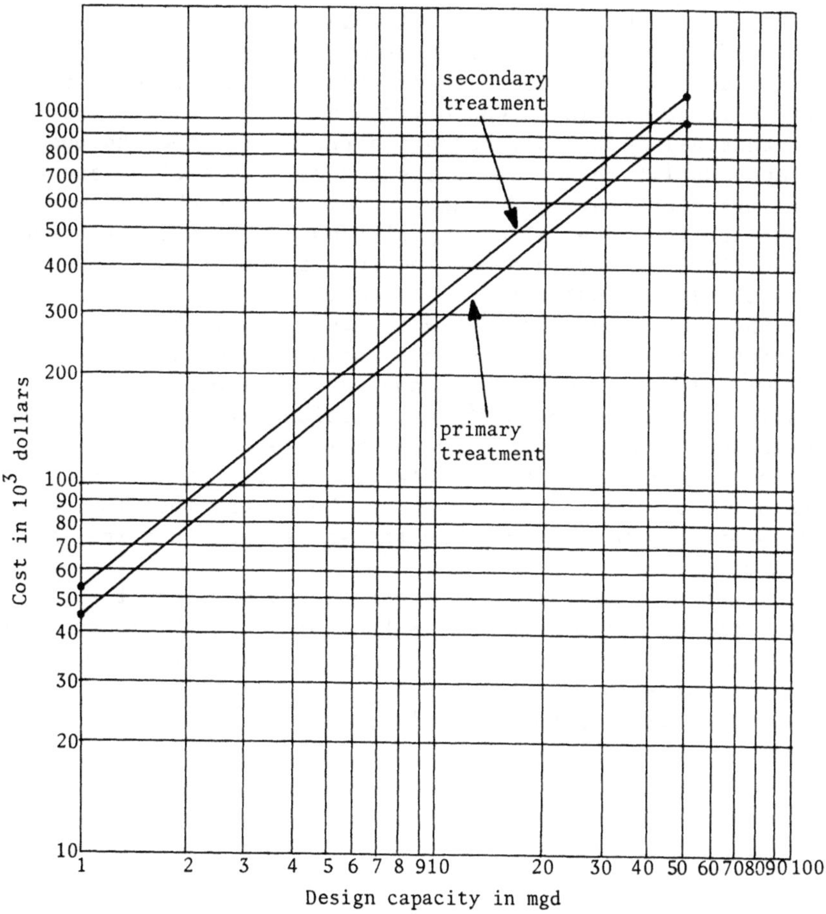

Figure 3. Construction costs of treatment plants, based on Equations (2) and (3).

It should be noted that long pipelines are not favored. This is because water pressure on sewer walls may cause some seepage into the porous soil around the pipe. In a long main such infiltration could be a significant factor in groundwater pollution. Incidently, the same concern is felt regarding any long pipeline; the Alaskan oil line is a striking example of a construction which is vulnerable to defects and to accident.

In considering outfalls of treatment plants, it will become clear that the bay sites are most troublesome in terms of water quality. Ocean outfalls would normally be preferable, since it can be reasonably argued that the ocean is an unlimited sink for contaminants.

Effluents discharged into the bay can affect the waters in a number of undesirable ways. Although the actual health risks and degradation of quality are hard to measure, useful surrogates are available to describe these adverse impacts. These would include total fecal coliforms (intestinal bacteria), organic nitrogen and phosphorous compounds (which act as nutrient sources for algae and other phytoplankton), and organic compounds which decay over time due to bacterial decompositions. The latter are known as biological oxygen demanding substances or BOD, for short. It is customary to regard BOD as a useful measure of the lack of health of a water body since the bacterial action which leads to its decay tends to deplete the level of dissolved oxygen (DO) in the water. The DO level is easily monitored for different portions of the bay and excessively small values tend to indicate a serious loss in water quality. By itself, however, it does not reveal anything about health hazards to people or the loss in species diversity or productivity in the water. It is solely an indicator.

Let us assume that effluent discharges from the treatment plants (also measured in tons of BOD/day) are continuous at a more or less steady rate. In practice, the flow can be somewhat variable but since we will be averaging over long time spans, such fluctuations are not significant.

3 Water Quality Model

The reason that the receiving waters of the bay shown in Figure 2 are susceptible to pollution from waste discharges while the ocean is relatively unaffected, is because the flushing action of the tides in the ocean is considerably more vigorous than is possible in the bay and the dispersion greater. In order to assess the impact of various waste treatment alternatives on the bay, we need to quantify what the long term pollutant concentrations in the bay will be as a result of tidal fluctuations. In the Long Island '208', an elaborate water quality model is utilized to give this information, based on the solution of partial differential equations of hydrodynamics in bays and estuaries. There are simpler versions that one can usefully employ, although these will generally be less accurate. For the purpose of the model described in the next section, one of these approaches will prove to be quite adequate. We begin with an intuitive discussion and later show how one can formalize the arguments by considering certain differential equations.

Let us then try to understand the basic idea behind the tidal model. To do this, consider a typical bay with an outlet to the ocean. We assume that pollutants discharged into the bay at some outfall are totally miscible with the water. Take, as a surrogate for the effluent, a dye which is injected into the bay at some point. The dispersion of the dye serves to simulate what actually happens to the pollutant. We will be interested in long term conditions in the bay, long after discharges have begun, when the pollutant concentrations have reached equilibrium with respect to the tidal movement. The approach used is tantamount to a steady state model in contrast with more elaborate versions which can follow short term, highly fluctuating, disturbances such as those caused by a storm surge.

When the dye is dumped into the bay, two things happen. First, because of movement of the water itself in each tidal cycle, some of the dye is carried seaward and eventually is lost in what is essentially an infinite ocean sink, never to return. Second, over several successive cycles, the pollutant plume folds back and forth on itself and becomes dispersed. Some additional dispersion occurs also because of the tendency for miscible substances to diffuse in the water (consider what happens when a dye is put into a full but perfectly still bathtub). Eventually the dye has spread throughout the bay and its concentration settles down to a value which varies only with location. This is the steady state condition referred to above. Of course, we assume that the dye is being continuously discharged into the bay at some outfall site for otherwise, with no further infusion, the concentration in the bay would ultimately be completely dissipated. Thus, steady state reflects a net balance between the amount injected into the bay and the dye lost to the ocean in each cycle after complete mixing has taken place.

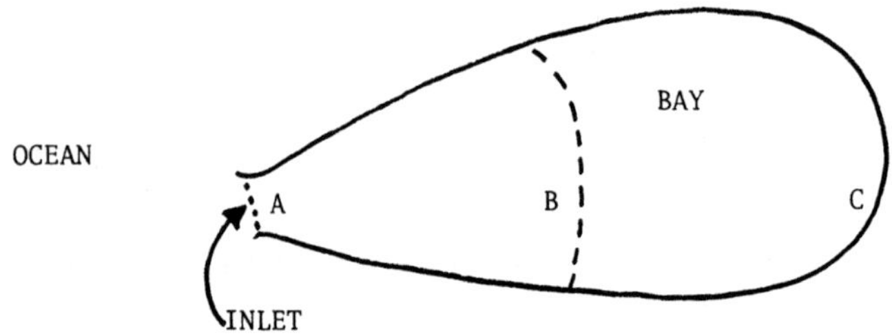

Figure 4. Schematic of a Hypothetical Bay.

Define a tidal excursion as the average distance travelled by the dye in each tidal cycle. Figure 4 is a crude schematic of a typical bay and we see indicated dotted contour lines, each of which represents a tidal excursion moving landward from the inlet at A. For instance, a dye injected at B would move to A during an ebb tide and as far as C during the ensuing flood tide. Now, suppose the discharge occurs somewhere within the bay at some given rate per day. Suppose the total volume of water crossing the boundary at A (the mouth of the bay in Fig. 4) is F_A (km^3 per day). Then in each tidal cycle roughly half of the flow is outward to the ocean and half is carried back towards land. This means that $F_A/2$ is the net flow which empties out. A total of W tons of dye also goes out to sea. It follows that

(1) $$c_1 = \frac{W}{F_A/2} = \frac{2W}{F_A}$$

is the average concentration of pollutant in the bay segment AB (in tons per km^3). Note that the pollutants flushed out during each ebb tide do not return since the flood tide brings back clean water. As we move one tidal excursion landward, the same thing is no longer true. In fact, in each flood tide, the segment BC receives some of the now partially polluted water from AB and therefore its steady state concentration, call it c_2, must be greater than c_1. To see this more concretely, let F_B be the total volume of water which crosses the boundary at B each day so that $F_B/2$ is the amount moving seaward with an equal amount moving back into BC from AB in each tidal cycle. If, as before, W represents the tons of effluent discharged into the bay, then the steady state concentration c_2 must equal the amount which is flushed out each cycle, given by $2W/F_B$, plus the amount which returns, which is $2W/F_A$. That is,

(2) $$c_2 = c_1 + \frac{2W}{F_B} .$$

Note, however, that this expression assumes a source discharge which originates in BC. If the outfall lies within AB, then the flow across B contributes pollutant in one direction only and so the term $2W/F_B$ is not applicable. In this case, c_2 is simply c_1:

(2') $$c_2 = c_1 .$$

In a bay consisting of more segments, the same intuitive arguments would be true and we can therefore appreciate that as one moves landward from an inlet to the head of the bay, one tidal excursion at a time, the pollutant concentrations will typically increase. The complete dispersion which characterizes steady state requires that all segments n landward of an outfall site have the same concentration as at the outfall location.

In order to obtain a convenient numerical measure of the values of the concentrations consider a 1 km^3 block of bay water which moves with the tide. If, in the steady state, a ton of pollutant is fully mixed into that volume then, since one km^3 of water weighs about 10^9 tons, it follows that the pollutant concentration is 10^{-9} (one part per billion or ppb, for short) in each such block. Depending on the total volume of bay water which is flushed out to sea (km^3/day) and on the effluent discharge rate W (tons/day) the concentrations will of course vary. Bays with poor tidal circulation can be expected to have higher than normal pollutant concentrations. In any event, these numbers are computable. For later reference, it is useful to note that, since one liter of water weighs 10^3 grams, then one milligram of pollutant in a liter represents approximately one part per million. It follows that 1 ppb = 10^{-3} mg/liter and 1 ppm = 1 mg/liter, where ppm means "parts per million."

Assuming a continuous discharge of one ton per day, the steady state pollution susceptibility at some point in a bay is defined as the long term concentration in ppb which occurs there. It may be of interest to know that for Long Island coastal waters these indices have been fully charted along contours of constant susceptibility. The resulting plots are similar to those displayed in Figure 5 for the sample region being discussed in this monograph. Thus, for instance, if the discharge site is at outfall location 6, then the resulting concentration due to one ton of pollutant entering the bay at this point each day is 300 ppb along a contour which stretches across the bay. The same steady state concentration holds at all parts of the bay to the right of this contour, as we know it must from the discussion above. On the other hand, the increased flushing nearer the inlet gives smaller pollutant values; the concentration along the contour which touches at outfall 5 is only 100 ppb. If the discharge is at outfall 4, then the resulting concentration is substantially increased to 1000 ppb, and between sites 6 and 4 the values increase continuously from 300 to 1000 ppb (shown is 500 ppb along some intermediate contour). In the event that the actual pollutant discharge is W tons per day, then all the above values are multiplied by W.

Let us introduce some convenient notation to use with the model to be discussed in the next section. The concen-

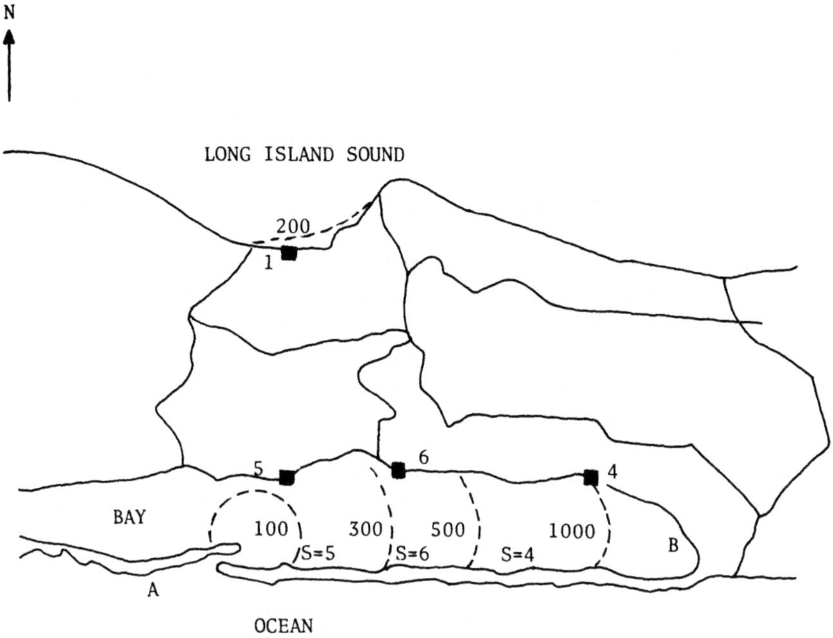

Figure 5. Contour Lines of Pollution Susceptibility in the Bay and Sound (ppb).

tration at contour s (shown in Fig. 5) due to one ton entering the bay at j, is given by values t_{js} in the following table:

		s: 5	6	4
j	5	100	100	100
	6	100	300	300
	4	100	300	1000

If the bay is fed by a significant river discharge, as would be the case in an estuary, then the resulting concentration would of course be less since the total flow towards the ocean is now greater. Also, if the pollutant is not conservative, in the sense that it decays in the water over time (such as BOD substances), then the long term concentrations, as given here would be somewhat inflated. Notwithstanding these reservations, it should be pointed out that the rather simple scheme outlined above (due to the oceanographer Weyl, Ref. 3) appears to give results which agree well with observed values.

Although the scheme presented above for estimating water quality in a bay has the virtue of simplicity, it is perhaps too heuristic for some tastes. Accordingly, we digress to present a somewhat more mathematical model, even if still quite intuitive, which leads to the same results.

Let us begin with the simplest case in which there is a flow in the longitudinal direction of a given water body. This means that a discharged substance from a waste treatment plant is not completely mixed in this direction. However, we do assume homogeneity laterally and vertically. The prototype of this situation is a river.

Consider now a slice of the water body (or river, if you prefer) as shown in Figure 6 in which Q is the flow (volume/time) and c is the concentration of the pollutant (mass/volume). The total volume of the slice is $V = A\Delta x$. It is convenient to assume that Q and A are constant in the variable x.

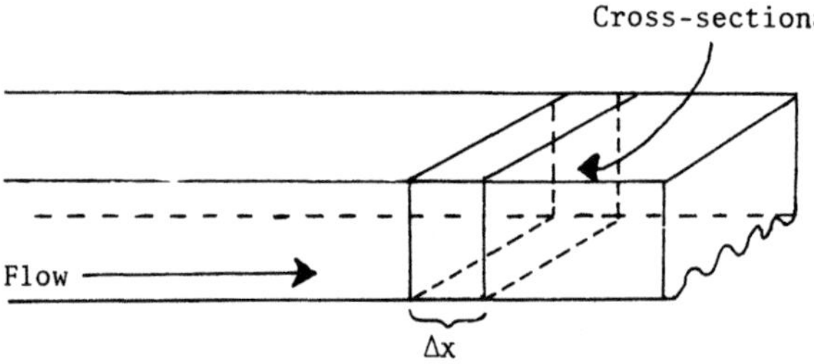

Figure 6.

During Δt, the total mass of the constituent entering the upstream face of the thin slice is $Qc(x,t)\Delta t$ and the mass leaving the downstream face is $Qc(x+\Delta x,t)\Delta t$. From Taylor's Theorem in the calculus, we know that

$$(3) \qquad c(x+\Delta x,t) = c(x,t) + \frac{\partial c}{\partial x}(x,t) \cdot \Delta x + o(\Delta x)$$

where $o(\Delta x)$ designates terms of higher order in Δx.

The total change of mass in the slice during Δt equals the difference in the mass of pollutant between what enters and leaves the slice plus any changes, accretions or losses, which occur within the slice itself. Let us momentarily ignore these accretions and losses or, as they are more commonly referred to, sources and sinks. Then the change in pollutant mass during Δt is $Q(c(x,t) - c(x+\Delta x,t))\Delta t$ which, for small Δt and Δx is approximately the same as

$V(c(x,t+\Delta t) - c(x,t))$. Equating these two quantities and using (3) gives, after dividing by Δt,

$$(4) \qquad V\cdot\left[\frac{(c(x,t+\Delta t)-c(x,t))}{\Delta t}\right] = -Q\cdot\left[\frac{\partial c(x,t)}{\partial x}\cdot\Delta x+o(\Delta x)\right]$$

Now divide by A x (note that $V = A\Delta x$) and let Δx and Δt both tend to zero to obtain

$$(5) \qquad \frac{\partial c}{\partial t} = -\frac{Q}{A}\frac{\partial c}{\partial x}.$$

If the pollutant concentration does not vary over time, then (3) and (4) shows that the inflow to the slice must equal the outflow which is simply the principle of conservation of mass.

Sources would include point sources such as waste effluent discharges from a treatment plant and distributed sources along the river bank from storm runoff and rainfall, while sinks include water pumping for irrigation and normal evaporation into the atmosphere. We will consider solely the point sources from sewage outfalls. However, instead of introducing these sources explicitly by the addition of a term in the equation, it is more convenient to bring in the effect of a source by means of a boundary condition $c(0,t)$ for $t \geq 0$. This will be done below. As for sinks, the only one to be considered is the loss of an organic pollutant through bacterial decay (as in the case of BOD). Assume that <u>the rate of decay is proportional to the remaining concentration at any time</u>. This rate is therefore expressed by the sink (loss) term $-kc$, where k is a constant of proportionality. When the pollutant is conservative, k is taken to be zero. Incorporating the loss term into (5) results in the modified equation:

$$(6) \qquad \frac{\partial c}{\partial t} = -\frac{Q}{A}\frac{\partial c}{\partial x} - kc.$$

This partial differential equation describes what is called the <u>advective model</u>, in which the quantity Q/A defines the <u>river velocity</u> in the longitudinal direction. Let us impose the <u>boundary condition</u>.

$$(7) \qquad c(0,t) = \frac{W}{Q}U(t), \qquad t \geq 0$$

$$\text{where } U(t) = \begin{cases} 1 & \text{if } t \geq 0 \\ 0 & \text{if } t < 0 \end{cases}$$

and W is the waste discharge rate (mass/time) from a sewage outfall. Then the solution to the first order partial differential equation (6) is given by (see Exercise 1):

$$c(x,t) = \frac{W}{Q} U(t-\frac{x}{u}) e^{-\frac{kx}{u}} \tag{8}$$

where u is the velocity Q/A. The quantity x/u is the time it takes for the effluent discharged at site zero to reach a distance x downstream. The concentration at x is zero for all time less than x/u and constant thereafter.

Numerical methods are expedient in the solution of (6). This is especially true when we wish to study water quality in an estuary or a bay where a one dimensional longitudinal flow is no longer an adequate representation. In this case the water body is partitioned into adjacent volume elements. Flow is taken to be directed into and out of each element along channels joining the centroids of the individual volumes. The quantity defined by c is assumed to be uniformly mixed within each element and all changes take place only between elements, not within them. Perhaps the best way to show how to proceed is to again consider the explicit example of the hypothetical bay illustrated earlier in Figure 4. Assume, as before, that the pollutant is conservative so that k = 0 and therefore one need only approximate the simpler equation (5).

This approximation is carried out by considering separately each of the two segments of the bay shown in Figure 4. Let Δx be the length of a tidal excursion from A to B or B to C and let ΔT represent the ebb-flow period (about six hours) outward towards the ocean. The flow across the mouth of the bay at A during ΔT is Q_1, and the flow across the contour at B is Q_2. Let c_i be the steady state pollutant concentration in segment i and suppose a waste treatment plant is discharging at rate W into segment 2 at location C.

By referring back to the way relation (4) was derived, we see that the net flow of pollutant across each volume element of longitudinal length in the direction from right to left (ebb-tide), during the tidal period of duration ΔT, can be approximated in segment 1 by

(9)

$$V_1 \frac{\Delta c_1}{\Delta T} = -(Q_1 c_1 - Q_2 c_2) \text{ while in segment 2,}$$

$$V_2 \frac{\Delta c_2}{\Delta T} = -(Q_2 c_2 - Q_2 \frac{W}{Q_2}) = -Q_2 c_2 + W$$

where V_i is the volume of segment i, Δc_i is the change in c_i over a tidal cycle, and W/Q_2 is the concentration of pollutant discharge into segment 2. Similarly, during a flood-tide of duration ΔT in which the flow is from left to right into the bay, there is a corresponding set of equations relating net flow across the elements:

$$(10) \qquad V_1 \frac{\Delta c_1}{\Delta T} = -Q_2 c_1$$

$$V_2 \frac{\Delta c_2}{\Delta T} = Q_2 c_1 + W.$$

Observe that in the first of equations (10) there is no pollutant flow into segment 1 across the boundary at A since the ocean returns only clean water in each flood-tide. This is also reflected in the first of equations (9) in which $-Q_1 c_1$ is a loss of pollutant into the open sea never to return. In both sets of equations, the term W can be thought of as a source term which is added to the equation (5).

In a steady state we expect Δc_1 over a flood-tide to be equal in value but opposite in sign to Δc_i over an ebb-tide, for each i. Employing this fact gives us the following pair of equations, utilizing (9) and (10):

$$(11) \qquad Q_2 c_2 - Q_1 c_1 = Q_2 c_1$$

$$-Q_2 c_2 + W = -Q_2 c_1 - W.$$

From the second of these equations, we find

$$(12) \qquad c_2 = c_1 + \frac{2W}{Q_2},$$

and substituting into the first relation of (11) yields

$$(13) \qquad c_1 = \frac{2W}{Q_1}.$$

Interpretating Q_1 as F_A and Q_2 as F_B makes it plain that (12) and (13) are identical to relations (1) and (2) obtained earlier by more intuitive but less formal arguments. Chacun a son goût.

Exercises:

1. Show that (8) is the solution to (6) with boundary condition (7). Hint: Let

$$c(x,t) = v(x,t)e^{-\frac{kx}{u}}.$$

Substituting into (6) gives a homogeneous partial differential equation. Now let $\zeta = t - x/u$ and attempt to prove that $v(x,t)$ is of the form $f(\zeta)$ for some function f.

2. A further extension of the advective model allows for diffusion. In estuaries, where there is back and forth tidal motion, there is also dispersion due to velocity shears and mixing due to salinity gradients at the salt and fresh water interface. In a river, for example, diffusion gives rise to the differences displayed in the figure below for a dye injected uniformly across the lateral direction.

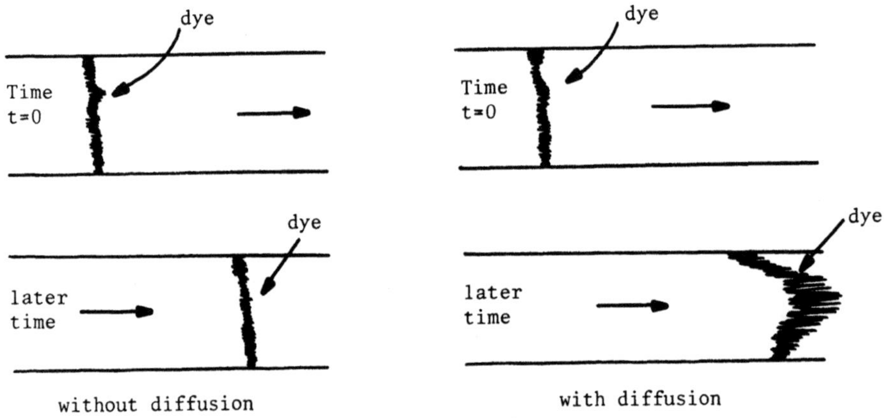

Refer again to the thin slice of a river of width Δx pictured earlier in Figure 6. The amount of diffusion across this slice in time Δt is proportional to the difference in the concentration c of the pollutant at each face. The greater the difference the more rapid is the diffusion. The mass of the constituent entering the upstream face in Δt, due to diffusion, is

$$- EA \frac{\partial c}{\partial x} \Delta t$$

where E (area/time) is a dispersion coefficient which is determined empirically by measurements in the actual water body. It is assumed to be a known constant. The minus sign indicates that diffusion occurs in the direction of decreasing concentration. As before, lateral and vertical homogeneity are assumed.

The mass of the substance which leaves the downstream face of the slice, as a result of diffusion, is derived in a way which is similar to that used earlier for advective flow in Equation (3). In fact Taylor's theorem gives

$$\frac{\partial c(x+\Delta x,t)}{\partial x} = \frac{\partial c(x,t)}{\partial x} + \frac{\partial^2 c(x,t)}{\partial x^2} \cdot \Delta x + o(\Delta x).$$

From this expression derive an analog of relation (4) and show that

$$\frac{\partial c}{\partial t} = -\frac{Q}{A}\frac{\partial c}{\partial x} + E\frac{\partial^2 c}{\partial x^2}. \tag{14}$$

Equation (14) is similar to Equation (5) except that a diffusion term has been added to the advective one. It can also be viewed as the familiar heat equation with an advective term added.

3. Consider again the bay shown in Figure 4. Suppose now that there is a pollutant discharge from a treatment plant at rate W into segment 1 at site B and no point source discharge into segment 2. Find the analog of Equations (9), (10), (11) and solve for c_1 and c_2. Do these values coincide with those obtained earlier in Equations (1) and (2')?

---

4 The Siting Model

With all preliminary considerations out of the way we can get down to the siting model itself, following the layout exhibited in Figure 2.

Let the service districts be indexed by i (i = 1, ..., 5) and the plant locations by j (j = 1, 4, 5). Both Districts 1 and 5 are to be sewered. Two options exist in each zone: either all waste flow generated in district i is treated at the plant i or all of it is diverted by pumping it to plant j where it is then treated and discharged (i = 1,5 and j = 5,1). This is indicated in Figure 2 by a dotted line connecting the two plants. Also shown are the outfall locations. It is assumed that maximum population density has already been achieved in Districts 1 and 5 so that future growth will occur only in the remaining zones.

District 2 is of sufficiently low density that no sewering is required. It is an affluent residential area which has effectively precluded future development through stringent zoning restrictions. District 4 requires sewering. Since the outfall site of plant 4 (which absorbs the flow of this zone) is in a part of the bay having high pollution susceptibility, a second outfall, located at site 6, is contemplated. Part or all of the flow to site 4 may be diverted to 6 as shown by the dotted line.

Sewers are also required in District 3. The effluent will be pumped along a trunk main to either of plants 1 or 4. Only one or the other of these alternatives is permitted. These possible connections are indicated by dotted lines in Figure 2.

Let q_i denote the total daily flow of sewage (mgd) originating in district i. It is obtained by multiplying the population in i by the 100 gallons/day generated by each person. For i = 1, 3, 4, 5 we assume it is known that the q_i are given by 12.2, 45.4, 9.2, and 31.2 mgd, respectively.

Denote by $x_{ij} \geq 0$ the amount of wastewater in mgd sent from district i to disposal facility j. Then it is apparent that one must have

$$\begin{aligned} q_1 &= x_{11} + x_{15} \\ q_3 &= x_{31} + x_{34} \\ q_4 &= x_{44} \\ q_5 &= x_{51} + x_{55} \end{aligned} \tag{15}$$

with all other x_{ij} values being zero. Now we need some decision variables. Let

$$y_j = \begin{cases} 1 & \text{if plant } j \text{ is built} \\ 0 & \text{otherwise} \end{cases}$$

and define the total daily flow (in mgd) into plants j (j = 1, 4, 5) by δ_j. If δ_6 is the diversion from plant 4 to the outfall at site 6, then

$$\begin{aligned} \delta_1 &= x_{11} + x_{31} + x_{51} \\ \delta_4 &= x_{34} + x_{44} - \delta_6 \\ \delta_5 &= x_{15} + x_{55} \end{aligned} \tag{16}$$

we also impose the reasonable condition

$$0 \leq \delta_6 \leq x_{34} + x_{44}. \tag{17}$$

In order to obtain a feasible solution, at least one of the plants 1,5 must be constructed:

$$y_1 + y_5 \geq 1. \tag{18}$$

Since plant 4 needs to be built to guarantee a feasible solution,

$$y_4 = 1. \tag{19}$$

Moreover, it is necessary to stipulate that if either of treatment plants 1,5 is not constructed, then no flow can be sent there, while if both are then all waste flow generated within district i must go to plant i (the last requirement insures that each treatment facility service its own local district). These are stated mathematically in the following way.

$$
(20) \qquad
\begin{aligned}
x_{31} + x_{51} &\leq (q_3 + q_5) y_1 \\
x_{15} &\leq q_1 \; y_5 \\
x_{11} &= q_1 \; y_1 \\
x_{55} &= q_5 \; y_5
\end{aligned}
$$

In the event that plant 1 is built, we need to constrain the flow of District 3 to a choice of plant 1 or plant 4, one way or the other. This is expressed as

$$
(21) \qquad x_{31}(q_3 - x_{31}) = 0
$$

since this forces either x_{31} or x_{34} to be zero. Relation (21) is a nonlinear statement of preference for one alternative over another.

In order to also have an effective tool for pollution abatement, one would want to minimize the impact of effluents discharged into the bay. To keep the model reasonable in size, we limit ourselves to considering only one undesirable aspect of effluent, namely BOD (see discussion in Section 3). We already remarked that the movement of the tides provides the flushing action necessary to cleanse the bay. As one moves inland from the bay inlet at A in Figure 5 to point B, there is a general increase in water pollution due to a continuous discharge of effluents, since the dispersive action of the tides becomes less effective.

In Section 3 we established the concentration t_{js} of pollution at s due to one ton of BOD entering at j (Figure 5).

We can now quantify pollution load in the bay resulting from a given configuration of facilities. It is known that a rough average of .83 tons of BOD are generated for each mgd of source wastewater. Primary treatment at a plant removes 30% of the BOD, while secondary treatment removes 85% of BOD and therefore leaves 15%. Since δ_j is the daily throughput at facility j, then 0.12 δ_j tons per day of pollutants are discharged into the bay or Long Island Sound, using secondary treatment. Therefore,

$$
\beta_j = .12 \; \delta_j
$$

is the tons per day of effluent entering the bay at site j for j = 4, 5, 6, while $\beta_1 = .12 \; \delta_1$ enters the north shore surface waters. It follows that the BOD concentration at location s in the bay from a source discharge at j is given by $\beta_j t_{js}$ in ppb. If we sum over all s we can gauge an average pollution loading in the bay due to source discharged from each j by

$$P = \frac{1}{3} \sum_{j=4}^{6} \sum_{s=4}^{6} \beta_j t_{js}.$$

We might wish instead to measure pollution at some specific location s. In this event P is replaced by

$$P_s = \sum_{j=4}^{6} \beta_j t_{js}.$$

It remains to discuss cost. From the conclusions reached in Section 2, the annual costs of the plants, in thousands of dollars, is given by

$$f_j = 52\ \delta_j^{0.8}$$

for j = 1, 4, 5. The cost of the trunk lines between sewer districts are

$$c_{ij} = 1.8\ x_{ij}^{0.6}$$

multiplied by the mileage from i to j (thousands of dollars per mile). Distances in miles between sites are provided by the following table:

	5	1	4	6
1	11			
5		11		
3		6.5	4.1	
4				5

Therefore the c_{ij} coefficients may be written as

$$c_{15} = 2.8\ x_{15}^{0.6} \cdot 11 = 30.8\ x_{15}^{0.6}$$

$$c_{51} = 2.8\ x_{51}^{0.6} \cdot 11 = 30.8\ x_{51}^{0.6}$$

$$c_{31} = 2.8\ x_{34}^{0.6} \cdot 6.5 = 18.2\ x_{31}^{0.6}$$

$$c_{34} = 2.8\ x_{34}^{0.6} \cdot 4.1 = 11.5\ x_{34}^{0.6}$$

$$c_{46} = 2.8\ \delta_6^{0.6} \cdot 5 = 14\ \delta_6^{0.6}\ .$$

At this point we have that

Dollar cost = plant cost + diversion cost.

Pollution cost = BOD from plant sewage.

The object of the siting model is to design a configuration of disposal facilities so that dollar cost and pollution cost are both as small as possible. Of course, one recognizes immediately that these are two conflicting objectives and so we resort to the usual expedient of treating one of the objectives as a constraint. Here pollution level is chosen as a constraint by writing

$$P_S \leq \text{constant}. \tag{22}$$

As the constant in (22) is varied, it will be possible for us to obtain a tradeoff curve between the two goals.

In order to formulate our problem as a mathematical program, let us first write down the expression for total dollar cost, D (in 10^3 dollars) as

$$\begin{aligned} D = 52(\delta_1^{0.8} + \delta_4^{0.8} + \delta_5^{0.8}) + 30.8\, x_{15}^{0.6} + 30.8\, x_{51}^{0.6} \\ + 18.2\, x_{31}^{0.6} + 11.5\, x_{34}^{0.6} + 14\, \delta_6^{0.6} \end{aligned} \tag{23}$$

(24) The problem is to minimize D subject to (15), (16), (17), (18), (19), (20), (21) and (22) in the variables $x_{ij} \geq 0$, the zero/one variables y_1, y_5 and δ_6.

The model, as presented, captures some of the tradeoff questions one would like to ask in our Long Island study regarding the siting of sewage disposal facilities. However, there are a number of other factors which the model simplifies or disregards which must be treated in an adhoc basis as a complement to the numerical results one obtains. These are discussed in the next section.

5 How The Siting Model is Used

Having formulated the model we would now want to address ourselves to its implementation. This is not as straightforward as it may appear. To begin with, the engineering consultant on the '208' project may be more interested in enumerating some specific options which are feasible from the viewpoint of cost and reliability and less concerned with selecting a single 'best'. Other considerations could intervene in the selection of alternatives which may well sidetrack the optimal model choice. Also, the model, as we formulated it, does not capture water quality indicators except for BOD even though in some sectors other considerations may predominate. In beach areas, for instance, total fecal coliforms would be viewed as a more insidious threat to public health.

Even more significant, perhaps, is the fact that every proposed 'solution' must be more or less able to overcome certain additional barriers which were not formulated within the model. It is not enough to have a plant configuration which is reliable and not too costly if it cannot be financed or if it does not satisfy certain legal and regulatory requirements. There are federal effluent standards which may need to be met, an adequate fiscal base must exist in the region to support the annual maintenance costs, sewage linkages which straddle several townships need to satisfy local zoning ordinances, questions of jurisdiction must be resolved, and so on. Most important could be the reaction of local citizen groups, each of which represents interests which do not always coincide. Property owners are concerned not only with cost but with the disruption caused by sewer construction and they would voice opposition to the choice of a specific location for a treatment plant in "their backyard." Clam fishermen may be in opposition to boat owners each of which wish to protect different parts of the bay. The local water authority which favors recharge options may be at odds with the public health agencies who fear that even treated wastes will ultimately degrade the quality of the underground supply of drinking water. Trade unions might favor the boost to the local economy which would result from a massive sewer construction project, whereas taxpayers groups would likely resent the extra tax burden. In short, public acceptability, together with many other factors, contributes significantly to what the planning board eventually adopts in the way of a waste treatment plan.

In practice the engineering consultants will select a few of the most reasonable alternatives and present them to the planning board and to various citizen groups for review. The final selection will more than likely not be made on technical grounds alone, as we saw. What is of interest to us here is the way the initial choice of options is determined. Obtaining explicit solutions from the model for different values of the water quality parameter P_S is one possibility but a more likely approach is one which narrows down the alternatives to a manageable few and then explicitly enumerates those which remain. This weeding out of candidates is based on heuristic judgements which conform to conventional engineering wisdom about what will work and what won't. By returning to our model once again, we can exercise our own judgment in a similar way. Consider, then, plant 1. Since the shoreline in the vicinity to the outfall is a productive shellfish area as well as a wildlife refuge, it would be preferable to divert all wastes to District 5. There are two reasons for not doing this. First the pipeline would have to go through the most populated area of the two districts resulting in substantial disruption and inconvenience. Second, we want to leave open the possibility that plant 1 will treat the sewage from District 3. There is a tradeoff here between discharging these wastes into Long Island Sound or in the bay and this should be left for the model to screen. If it is decided to construct a major facility at site 1, than an ex-post-facto decision can be made to extend the outfall further out into the Sound. However, this is not explicitly treated by the model. Let's now take a look at plant 5. It has already been decided that a hookup to plant 1 is not desirable. When this is coupled to the fact that the proposed bay outfall is in a zone of low pollution susceptibility, we have argued in favor of constructing plant 5. As a result both of the Districts 1, 5 will have waste treatment facilities. This leaves us with two options. The first is when District 3 connects to 1 and the other is when it connects to 4. Finally, consider District 4.

When site 4 diverts part of its flow to plant 6, an important tradeoff occurs. The outfall of 6 is in an area of lower pollution susceptibility so the more that is diverted, the less the pollution. However, cost increases for construction of the trunk line joining 4 to 6. We will examine the tradeoffs below.

Taken together, the stipulations given above narrow down the choices available in the model to a total of essentially two options. They are cataloged below:

1. Build plants 1 and 5. District 3 disposal is at plant 1.

2. Same as in 1 except that District 3 disposes at plant 4.

It is quite apparent that with only two alternatives an explicit enumeration of the costs of each, followed by a simple comparison to determine the best option, is carried out more simply than by attempting to compute an optimum value from the nonlinear mixed integer model derived in the previous section. (Such models are known as "fixed-charge plant location.") Anyone with any experience in coding such programs knows that the task can be disagreeable and time consuming, whereas with the aid of a desk calculator, comparison of two options is relatively routine. The explicit computations are left to the reader and the end results are summarized in Table 1. Please note that we have chosen to monitor pollution at only one site in the bay where tidal flushing is least effective (s = 4). In this case

$$P_4 = \sum_{j=4}^{6} \beta_j t_{j4}.$$

Also computed is pollution in the north shore using

$$P_1 = 200 \beta_1.$$

TABLE 1

	Daily sewage generated (mgd)	Daily outflow from plant i in district (mgd)	BOD from sewage treatment plant i (ppm)	Total BOD concentration from all sources (ppm)	Total structural cost in annual dollars
Option 1					$2,633,320
Districts					
1	12.2	57.6	1.38 (North Shore)	1.69 (North Shore)	
3	45.4				
4	9.2	9.2	1.1 (s = 4)	1.89 (s = 4)	
5	31.2	13.2	.37 (s = 5)		
Option 2					$2,589,200
Districts					
1	12.2	12.2	.29 (North Shore)	.60 (North Shore)	
3	45.4				
4	9.2	54.6	6.55 (s = 4)	7.34 (s = 4)	
5	31.2	31.2	.37 (s = 5)		

By looking at the Table, it is apparent that the cheapest alternative is the second. Since BOD concentrations at s = 4 do not contravene the generally accepted water quality standard of 5 ppm dissolved oxygen, it seems like the option to choose. (The saturation value of dissolved oxygen (DO) is roughly 9 ppm so that if we subtract BOD concentrations from 9, this gives an estimate of DO. Note that we are now measuring in units of ppm, not ppb.) And, so it is done. Option 1 is not picked because of strong opposition from the politically effective north shore communities since they claim that it would increase pollution by nearly 300% (which is true). Besides, under option 2 one can consider diverting part of the flow to a second outfall at site 6 in order to reduce excessive BOD concentrations in the bay. We therefore turn our attention to this now, by computing total dollar cost versus total BOD at site s = 4 from all sources (including the outfall from plant 6). The results are given in Table 2 below and plotted in Figure 7. We see that for a modest increase in cost there is a substantial reduction in pollution loading at the most sensitive portion of the bay if all sewage is diverted to the outfall at 6.

As a result of this analysis, the Planning Board would likely recommend a single secondary treatment plant with outfall at site 6 of capacity not less than 54.6 mgd (to service districts 3 and 4).

TABLE 2

Daily Outflow at Plant 4 (mgd)	Daily Outflow at Plant 6 (mgd)	BOD from all Sources at site s = 4 (ppm)	Total Annual Cost of Configuration (Dollars)
54.6	0	7.34	2,589,209
40	14.6	6.11	2,659,151
27	27.6	5.02	2,691,697
15	39.6	4.02	2,716,485
0	54.6	2.76	2,743,537

The problem of finding an optimal configuration of facilities and linkages between them can also be approached by use of a different model. We begin by enumerating the various options available for providing service to one or more of the districts. As the discussion in the previous section indicates, these reduce to four possibilities.

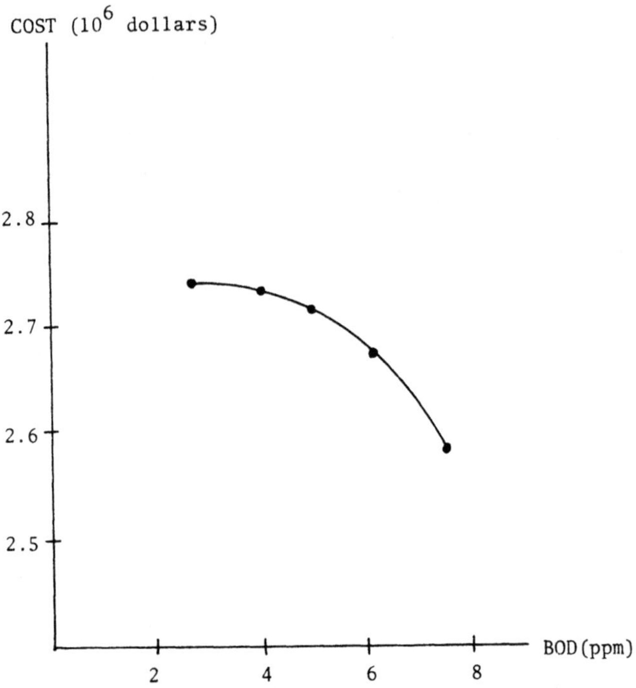

Figure 7. Cost of Diverting Sewage Versus Pollution (From Table 2).

Option one consists in building plants 1 and 5, with District 3 serviced by site 1, option two is the same except that District 3 is not handled by the plant at site 1, the third option is to build plant 4 with District 3 serviced at site 1, while the last option is to have plant 4 service District 3. Let a_{ik} be entries of a 4 by 4 matrix defined as

$$a_{ik} = \begin{cases} 1 & \text{if district } i \text{ is serviced by option } k \\ 0 & \text{if otherwise.} \end{cases}$$

Also let x_k be decision variables given by

$$x_k = \begin{cases} 1 & \text{if option } k \text{ is chosen} \\ 0 & \text{if otherwise.} \end{cases}$$

The cost of each option is readily computed and is denoted by c_k. What one seeks to do is choose $x_k \geq 0$ so as to minimize

$$\sum_{k=1}^{4} c_k x_k \qquad (25)$$

subject to

$$(26) \qquad \sum_{k=1}^{4} a_{ik}x_k = 1 \qquad \text{for} \quad i = 1, 3, 4, 5.$$

The significance of relation (26) is that it ensures that each district is covered exactly once by some sewage treatment plant. Relation (25) is a statement of minimal total cost.

Since each of the 4 alternatives service one or more districts, the model formulation is an example of what is known as <u>set covering</u>. A covering of a set S is a collection of not necessarily disjoint sets $\{S_i\}$ whose union contains S. If a "cost" is attached to each S_i, then one wishes to extract a subset of the collection whose total cost is as small as possible and whose union still contains ('covers') S. Such problems occur in a variety of settings of interest to the applied mathematician, most typically in the context of facility location and scheduling questions.

In order to solve the above problem, one requires an integer programming code. For problems of the size given here, simple codes are available at most computing centers (usually as part of larger mathematical programming packages). However, it sometimes happens that if one ignores the integer requirement and simply attempts to find a solution by linear programming, then the solution will turn out to be integer-valued anyway. Algorithms for finding integer solutions (usually by "branch and bound" methods) are more complicated than the more straightforward simplex algorithm of linear programming and so the latter is a preferable first approach. In our case, this is simple to do. The relations (26) can be written out as

$$\begin{aligned} x_1 + x_2 \qquad\quad &= 1 \\ x_1 \qquad\qquad + x_4 &= 1 \\ x_3 + x_4 &= 1 \\ x_1 + x_2 \qquad\quad &= 1. \end{aligned}$$

The fourth relation is redundant, and from the remaining we see that $x_2 = x_4$, $x_1 = x_3$. Thus if $x_1 = \alpha \geq 0$, then it is apparent that $\alpha \leq 1$. This allows the objective function (25) to be written as

$$(c_1+c_3-(c_2+c_4))\alpha + (c_2+c_4).$$

We see immediately that the minimum of this expression requires that α be zero or one, depending on whether or not c_1+c_3 is greater than or equal to c_2+c_4. In our case, c_1+c_3 is \$2,633,320 and c_2+c_4 = \$2,589,200 (Table 1). Therefore, $x_1 = x_3 = 0$ while $x_2 = x_4 = 1$. This gives the same solution as obtained from the fixed charge plant location model.

In the general case of an m by n matrix A with entries a_{ik} and n-vector x with coefficients x_k the set covering problem is to minimize $c^T x$ subject to $Ax = 1$ where c^T is the transpose of an n-vector consisting of non-negative costs for the various options. The optimal x is to be non-negative and integer valued.

Let us close this section with a brief comparison of the two models.

Which of the two to use is largely a matter of taste. There are some advantages and disadvantages to each. The set covering approach, although conceptually simpler than the plant location model, is burdened by the need to carry out an extensive pre-screening of options in order to formulate feasible clusters of possible sitings and linkages. In larger problems, this becomes cumbersome and one runs the risk of overlooking some potentially attractive alternatives by the "eyeballing" selection procedure. By contrast, in the other approach, decisions on what constitutes feasible linkages take place within the model itself. However, the plant location program generally involves considerably more constraint equations than for set covering. A more serious objection to the plant location program is that the objective function is not only more complicated but unless its individual terms are expressed as linear functions, the problem may be difficult to solve. In the set covering case, however, the cost computation is carried out prior to setting up the model, and the result is expressed as a single numerical coefficient c_k for each option. Although this computation of c_k may involve linear or non-linear cost considerations, the model itself is impervious to how the cost was arrived at.

Exercises:

4. Suppose the planning staff modifies the configuration of facilities to allow for the additional possibility of all wastes generated in District 4 being conveyed to plant 1 for disposal. In this case the plant in District 4 would not be constructed. The distance of a trunk line from District 4 to plant 1 is 13 miles.

Rewrite the constraining relations (15) through (22) to account for this new possibility and modify the cost (23) accordingly.

5. In the general set covering problem to minimize $c^T x$ subject to $Ax = 1$ and $x \geq 0$, x integer-valued, given $c \geq 0$, show that the optimal solution vector x must in fact have zero or one components. Assume that A consists of 0's and 1's.

---

6 Last Thoughts

It is important to note that the final configuration of facilities and of recommended zoning changes do not result in a regional plan of lowest possible pollution levels. It is in fact a compromise solution in the face of requirements which were not explicitly articulated by the model.

At this point some readers will be somewhat perplexed. After all, we invested some effort to develop a mathematical model only to be told that in practice intuitive insights usually suffice to enumerate a small number of options which can then be compared by simple routine computations. So what good is the model? Well, the answer rests with the observation that in order to formulate appropriate options which exhibit the interrelations of the several factors which come into play, one must implicitly or explicitly structure the problem so that the costs and constraints are clearly understood as part of a coherent whole. To some experienced practitioners it may suffice to see the geographic configuration of facilities and pipelines, as shown in Figure 2, in order to grasp the options. To others, the mathematical model conveniently lays out the possibilities. The model is, therefore, a tool for organizing the multitude of options, before other considerations are employed to narrow down the choices. Without the model, judgments tend to be ad hoc and incomplete and there is little chance of getting an overview of the possibilities. Moreover, the mathematical formulation makes explicit the quantities over which to optimize and the quantities which are traded off against each other. With this information, preference for one alternative over another can be clearly delineated and used as a basis for further discussion. The larger and more complex the problem, the more useful this is.

7 References

1. "Executive Summary of Section 208 Program for Designated Areas," EPA (Washington, D.C.), Oct., 1974.
2. "Long Island Comprehensive Waste-treatment Management Plan," 2 Volumes, Nassau-Suffolk Regional Planning Board (Hauppauge, NY), 1978.
3. Weyl, Peter, "Pollution Susceptibility: An Environmental Parameter for Coastal Zone Management," Coastal Zone Management Journal 2 (1976), 327-343.

A more comprehensive approach to models in which there are conflicting objectives is provided by:

4. Cohon, Jared, Multi-objective Programming and Planning, (Academic Press, NY), 1979.

A simple treatment of partial differential equations of the first order, such as equation (5), may be found in the ninth chapter of:

5. Apostle, Tom, Calculus, Volume 2 (2nd edition), John Wiley, 1969.

Fixed-charge plant location and set-covering models as well as integer programs in general, are usefully surveyed in:

6. Wagner, Harvey, Principles of Operations Research. (Prentice-Hall), 2nd edition, 1975.

Water quality models in rivers and estuaries are discussed more fully in the following book:

7. Thoman, Robert, Systems Analysis and Water Quality Management (McGraw-Hill), 1972.

APPENDIX A
Water Quality Management in a River

Consider the following problem. There is a river, represented by the x axis, in which waste treatment plants are located at sites $x = x_A \equiv 0$ and $x = x_B$. These pump treated sewage from neighboring townships into the river having an average flow velocity u. Downstream is in the positive x direction. The effluent consists of BOD in the amounts L_A and L_B at the two locations and by treatment there is a reduction by a fraction α of L_A and fraction β of L_B. Cost of treatment is assumed proportional to the percentage reduction (although in practice, these costs could be somewhat nonlinear). Governmental water quality standards prescribe a minimum dissolved oxygen level of D_{min} anywhere in the river. We want to achieve this objective at a minimum total cost of treatment. Formulate a model for water quality in the river and then use this to determine the least cost treatment policy. To make the problem interesting, assume that BOD decays over time even though our earlier discussions assumed BOD to be conservative. Also, assume that the water in the river upstream from site $X_A = 0$ is unpolluted and that effects of runoff from distributed sources along the river can be neglected. Also neglect the possibility of diffusion.

In order to make digestion of this problem easier, we proceed in stages through a series of exercises.

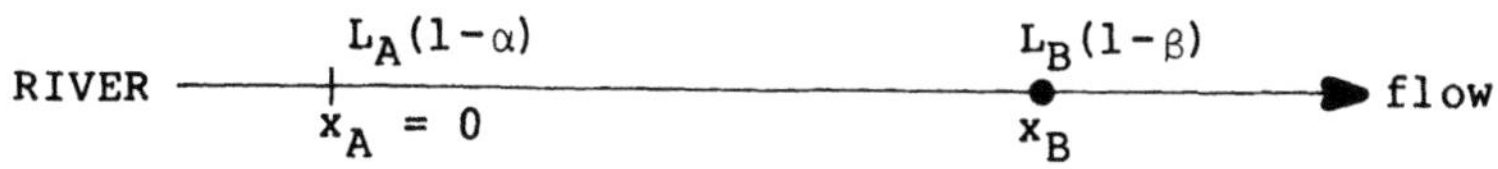

First of all it appears that the advective model of Equation (6) is suitable to describe the quantity of BOD at any point in the river. This equation takes into account the decay of BOD by oxidation through bacterial action. As it decays, it depletes the supply of dissolved oxygen, or DO. The rate of decay at any time is proportional to the amount of BOD remaining at that time, with the constant of proportionality written as K_L (in units of day^{-1}) and $L(x,t)$ denotes total BOD along the river (mg/liter). Equation (6) can therefore be written as

$$\frac{\partial L}{\partial t} = -u \frac{\partial L}{\partial x} - K_L L.$$

Now consider dissolved oxygen levels (mg/liter) along the river, denoted as $D(x,t)$. The rate of increase of D is negatively proportional to the amount of BOD decay (this is a "sink" term). There is also a "source" of DO due to the effect of reoxygenation (self-purification) of the river through surface to air contact. If D_S is the saturation (or maximum) level of oxygen which is possible in the water, then the rate of reoxygenation is taken to be proportional to the amount of oxygen deficit $D_S - D$ at any time. The constant of proportionality is K_D (in day^{-1}). We remark that although D_S is here assumed constant, in practice it depends somewhat on the temperature of the water.

Exercise

6. Derive a differential equation for D.

We now seek a steady state (time independent) solution to the coupled differential equations for L and D. Actually, it is more convenient to work with the oxygen deficit itself which we denote by $S = D_S - D$. Let $L(0)$ and $S(0)$ be the known initial values of L, S at the location $x_A = 0$ on the river. (Later we will consider location x_B). Then the solutions are

$$\text{(A.1)} \quad \begin{cases} L(x) = L(0)e^{-\frac{K_L x}{u}} \\[2ex] S(x) = \dfrac{K_L}{K_D - K_L}\left(e^{-\frac{K_L x}{u}} - e^{-\frac{K_D x}{u}}\right)L(0) + S(0)e^{-\frac{K_D x}{u}}. \end{cases}$$

Exercise

7. Show that (A.1) is the solution to the coupled differential equations for L, S in the steady state. The second of these equations, the one for S, is known as the Streeter-Phelps Equation, a classic in the water pollution literature. It was first formulated in a report issued in 1925 by the U.S. Public Health Service.

A typical profile of the dissolved oxygen deficit (also known as the oxygen "sag") is shown below. It is obtained from the Streeter-Phelps Equation. An initial loss of oxygen is eventually offset by the effects of reoxygenation as we move downstream. Notice that distance x along the river can be converted into units of time required to reach x by letting $t = x/u$.

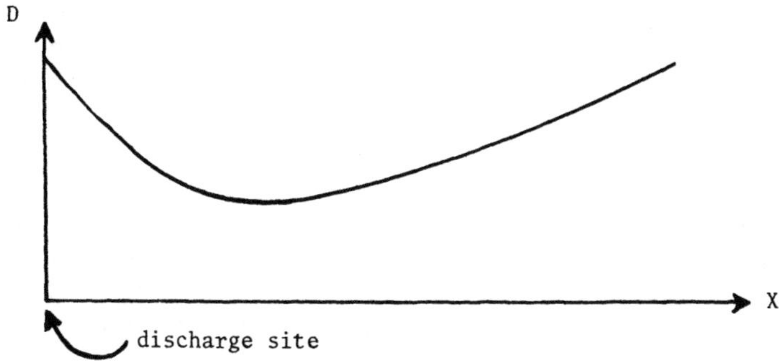

Exercise

8. Use the Streeter-Phelps Equation to determine the downstream location x^* along the river at which D achieves a minimum value. Use the fact that $S(0) = 0$. Restricting ourselves to the segment of the river from $x_A = 0$ to x_B, show that the minimum of D has the value

$$\left.\begin{array}{ll} D(x_B) & \text{if } x_B < x^* \\ D(x^*) & \text{if } x_B \geq x^* \end{array}\right\}$$

Since a fraction of α of the BOD discharged at x_A is removed by treatment, then the initial condition $L(0)$ can be written as $L_A(1-\alpha)$ while $D(0) = D_S$ by assumption. In the segment of the river between $x_A = 0$ and x_B, dissolved oxygen reaches the minimum level obtained in the last exercise. Assume, for simplicity, that $x^* > x_B$ so that this minimum is $D(x_B)$. When $x^* \le x_B$; the analysis is entirely analogous to that which follows and so it is not necessary to consider it explicitly here. The first of equations (A.1) permits the value of $L(x_B)$ to be determined. In addition, at x_B there is a new influx of BOD in the amount of $L_B(1-\beta)$. Therefore, the total BOD at x_B is $\tilde{L}(x_B) = L(x_B)+L_B\,(1-\beta)$.

<u>Exercise</u>

9. Show that there exist constants c_i for which

$$D(x_B) = c_1 + c_2\alpha$$

$$L(x_B) = c_3 + c_4\alpha + c_5\alpha.$$

Using the values of $D(x_B)$ and $\tilde{L}(x_B)$ as initial conditions, the equations (A.1) can be used again to compute the minimum level of oxygen D^* downstream from x_B. From exercises (8) and (9), it is easy to see that for suitable constants

$$D^* = c_6 + c_7\alpha + c_8\beta.$$

We are now in a position to pose an optimization problem for optimal waste treatment. Total treatment cost is $\lambda(\alpha+\beta)$ for some $\lambda > 0$ since cost is proportional to the fraction of BOD removed. Water quality requirements force the constraints

$$\text{(A.2)} \qquad \begin{cases} c_6 + c_7\alpha + c_8\beta \ge D_{min} \\ c_1 + c_2\alpha \ge D_{min}. \end{cases}$$

The first of these inequalities is a statement of minimal water quality standards in that part of the river from x_B downstream, while the second inequality does the same thing upstream from x_B. The problem is to choose $0 \le \alpha \le 1$, $0 \le \beta \le 1$ so as to minimize the cost subject to (A.2). This is a <u>linear programming problem</u> of a very simple type. It is fairly evident that α, β are to be chosen to be as small as possible consistent with (A.2). For numerical purposes, it is useful to know that typical values of the equation parameters are K_D = .7/day, K_L = .3/day, D_S = 9 mg/liter, D_{min} = 5 mg/liter, u = 12 miles/day.

Exercise

10. If the constants c_i in (A.2) are positive and if $D_{min} > \max(c_1, c_6)$ show that at least one of the inequalities in (A.2) is actually an equality. This implies that along some portion of the river water quality standards are just barely met in an optimal solution.

APPENDIX B
Water Quality Management Through Zoning

In the Siting Model considered in Section 4, it was assumed that Districts 3 and 4 would both be sewered. We now relax this requirement by stipulating that this need not be the case for District 3, even though District 4 continues to be sewered as before. It is generally not economical to sewer an area in which the residential density is too low. On the other hand, since zones without sewers discharge wastes directly to ground through cesspools, there is concern that too high a density will result in a contaminated underground water supply. The engineering consultants choose a figure of <u>five households per acre</u> as a maximum residential density which is permissible without sewers in District 3.

Future residential development is taken to be one of the two types: low density, meaning an average of about one household per acre, or high density, which averages about ten households per acre. Let u_{i1} be total acres of projected low-density housing in district i (i = 3,4) and u_{i2} the corresponding acreage for high-density housing. These are quantities to be determined and, once known, they will be enforced by zoning restrictions in the two districts. Taken together, these areas constitute a single political jurisdiction (township).

The land area A_i available for residential use is given by A_3 = 25,300 acres and A_4 = 13,600 acres for a total of 38,900 acres in the township. In order to encourage energy savings through a reduction in residential heating the town plan specifies that, in the future, a total of 13,000 acres should be of high-density ("clustered") development. This means that

$$\text{(B.1)} \quad \begin{cases} u_{32} + u_{42} = 13{,}000 \text{ acres} \\ u_{31} + u_{41} = 38{,}900 - 13{,}000 = 25{,}900 \text{ acres.} \end{cases}$$

A certain amount of land is already built on at present. This provides lower bounds for future development in District 3:

$$\text{(B.2)} \quad \begin{cases} u_{31} \geq 9{,}400 \text{ acres} \\ u_{32} \geq 2{,}500 \text{ acres} \end{cases}$$

The average density of development in this area of the township is evidently

$$\text{(B.3)} \quad z_3 = \frac{u_{31}+10u_{32}}{25{,}300} \text{ households/acre.}$$

If $z_3 \leq 5$, then District 3 requires no sewering and so the total cost of the wastewater disposal configuration is less than that computed in Section 5 of the text. However, there is a price to pay as will become apparent presently. First, however, observe that since the waste loads vary with population, the quantities q_3, q_4 which were assumed fixed in the text are now variable. Assuming an average of 3.5 persons per household, each household generates 350 gallons of waste-water per day. This is the information needed to do the following exercise.

Exercise

11. Find an expression for q_3, q_4 in mgd in terms of u_{i1} and u_{i2}, $i = 3,4$. Lable this as relation (B.4).

An interesting tradeoff now takes place. If high-density building in District 3 is discouraged in the future, then, as we observed above, total dollar cost for construction is decreased because sewering can be avoided there. However, the waste load into the surface waters of the bay adjacent to District 4 is now increased. This occurs because the impermeable paved surfaces (roads, parking areas, etc.) in the land which adjoins the coastline of the district promotes the runoff of BOD wastes during each rain storm. It is generally assumed that this paved area is proportional to population density and, therefore, as clustered residential development shifts from the interior of the township to the coastal areas of District 4, pollution in the bay can be expected to increase. The added pollution from runoff at location s = 4 in Figure 5 (only this region of high susceptibility is to be considered here) is estimated to be

$$\frac{15u_{41}+40u_{42}}{10^4} \text{ tons/day} \tag{B.5}$$

a quantity which is to be added to the pollution generated from treatment plant outfalls.

At this point a modified version of the siting model can be formulated which incorporates the residential zoning options represented by the variables u_{i1} and u_{i2} for i = 3,4. We do this in two steps:

Exercise

12. Express the fact that q_3 is zero if $z_3 \leq 5$ and equals 8.83 z_3 otherwise, by means of a constraint (or set of constraints) together with the requirement that q_3 be as small as possible. Hint: Consider the expressions for q_3 and z_3 given by (B.3) and (B.4). Label this constraint (s) as (B.6).

13. Examine each of the constraints (15) through (22) in turn and determine which needs to be changed and, also, express any additional relations that are required, keeping in mind that there are now additional (non-negative) variables u_{i1} and u_{i2} to consider. To simplify matters, pollution is to be measured only at the site s = 4 shown in Figure 5.

APPENDIX C
Solutions to Exercises

1. Let $c(x,t) = v(x,t)e^{-\frac{kx}{u}}$ and substitute into (6) to obtain

$$\frac{\partial v}{\partial t} = - u \frac{\partial v}{\partial x}. \tag{C.1}$$

The original non-homogeneous equation has been rendered homogeneous. Now let $\zeta = t - x/u$ and let

$$H(x,\zeta) = v(x,\zeta+\frac{x}{u}).$$

Then

$$\frac{\partial H}{\partial x} = \frac{\partial v}{\partial x} + \frac{1}{u}\frac{\partial u}{\partial t} = 0$$

by virtue of (C.1). Therefore, H is independent of x and so $H(x,\zeta) = f(\zeta)$ for some f. That is

$$v(x,t) = f(t-\frac{x}{u}) \text{ and hence}$$

$$c(x,t) = f(t-\frac{x}{u})e^{-\frac{kx}{u}} .$$

Conversely, it is easy to see that if $c(x,t)$ is of this form, then it must satisy equation (6). Now impose the boundary condition $c(0,t) = (W/Q)U(t)$ for $t \geq 0$. Then, since $c(0,t) = v(0,t)$, it follows that $v(x,t) = (W/Q)U(t - x/u)$ from which we obtain solution (8).

2. The mass of the substance which leaves the downhill face during Δt is

$$-EA\left(\frac{\partial c(x,t)}{\partial x} + \frac{\partial^2 c(x,t)}{\partial x^2}\Delta x + o(\Delta x)\right)\Delta t.$$

The change in pollutant mass in the slice during Δt is

$$\left[Q(c(x,t)-c(x+\Delta x,t)) - EA\left(\frac{\partial c(x,t)}{\partial x} - \frac{\partial c(x+\Delta x,t)}{\partial x}\right)\right]\Delta t.$$

which is the same as

$$\left\{-Q\left(\frac{\partial c(x,t)}{\partial x}\Delta x\right) + EA\frac{\partial^2 c(x,t)}{\partial x^2}\Delta x + o(\Delta x)\right\}\Delta t.$$

For small $\Delta x, \Delta t$, the above change in mass is approximately the same as $V(c(x,t+\Delta t) - c(x,t))$. Equating these quantities, dividing by Δt and by $V = A\Delta x$, and letting $\Delta t, \Delta x$ tend to zero, gives

$$\frac{\partial c}{\partial t} = -\frac{Q}{A}\frac{\partial c}{\partial x} + E\frac{\partial^2 c}{\partial x^2}.$$

Observe that a steady state solution to equation (14) can be obtained exactly because $\partial c/\partial t = 0$ implies that one has a linear second order ordinary differential equation:

$$E\frac{d^2 c}{dx^2} - \frac{Q}{A}\frac{dc}{dx} = 0.$$

Letting $dc/dx = v$ gives a linear first order equation in v. A unique solution is obtained by imposing a condition $dc/dx = 0$ at $x = 0$ (which implies no diffusion beyond the source of the river), in addition to $c(0) = W/Q$.

3. The equation corresponding to (9), (10), and (11) are:

<u>Ebb-tide</u>

$$V_1\frac{\Delta c_1}{\Delta t} = Q_2c_2 - Q_1c_1 + W$$

$$V_2\frac{\Delta c_2}{\Delta t} = -Q_2c_2.$$

<u>Flood-tide</u>

$$V_1\frac{\Delta c_1}{\Delta t} = -Q_2c_1 + W$$

$$V_2\frac{\Delta c_2}{\Delta t} = Q_2c_1.$$

In a steady state situation, Δc_i in a floodtide has an opposite sign to Δc_i in an ebb-tide for $i = 1,2$. Proceeding as in the text, this results in the relations (identical to (1) and (2) if $Q_1 = F_A$):

$$c_2 = c_1$$
$$c_1 = 2W/Q_1.$$

4. The new relations are indicated by (15') through (22')

$$(15')\quad \begin{cases} q_1 = x_{11} + x_{15} \\ q_3 = x_{31} + x_{34} \\ q_4 = x_{41} + x_{44} \\ q_5 = x_{51} + x_{55}. \end{cases}$$

$$(16')\quad \begin{cases} \delta_1 = x_{11} + x_{31} + x_{41} + x_{51} \\ \delta_4 = x_{34} + x_{44} - \delta_6 \\ \delta_5 = x_{15} + x_{55}. \end{cases}$$

Relation (17) remains the same. At least one of plants 1 or 4 must be built:

$$(18')\quad \begin{cases} y_1 + y_4 \geq 1 \\ y_1 + y_5 \geq 1. \end{cases}$$

Relation (19) eliminated since it is now incorporated into (18').

$$(20')\quad \begin{cases} x_{31} + x_{41} + x_{51} \leq (q_3+q_4+q_5)y_1 \\ x_{15} \leq q_1y_5 \\ x_{34} \leq q_3y_4 \\ x_{11} = q_1y_1 \\ x_{44} = q_4y_4 \\ x_{55} = q_5y_5 \end{cases}$$

Relations (21) and (22) remain the same.

Finally, in (23) add a cost term to correspond to a possible trunk line from Disrict 4 to Plant 1. Add the term

$$36.4x_{41}^{0.6}.$$

5. Let x^0 be the optimal solution vector to the set covering problem. That is, $x_1^0 = (x_1^0, \ldots, x_n^0)$ with $x_1^0 \geq 0$, being integers. Suppose $x_\ell^0 > 0$ for some ℓ and that in fact $x_\ell^0 > 1$. If i is such that $a_{i\ell} = 1$, then

$$\sum_{j=1}^{n} a_{ij}x_j > 1,$$

which contradicts the constraint $Ax = 1$. Therefore there is no row i for which $a_{i\ell}$ is non-zero. This means that the ℓth column of A is zero and so we can redefine x_ℓ^0 to be zero and still satisfy the constraint equations. Moreover, total cost has now been reduced below the value

$$\sum_{j=1}^{n} c_i x_i^0 .$$

This contradicts the assumed optimality of x^0 and it therefore follows that x^0 has zero or one components.

6. Modify equation (6) by adding a source term for reoxygenation in addition to a sink term to account for decay of BOD since this results in an uptake of oxygen. Thus,

$$\frac{\partial D}{\partial t} = - u \frac{\partial D}{\partial x} - K_L L + K_D(D_S - D). \tag{C.2}$$

7. In a steady state $\partial L/\partial t = \partial D/\partial t = 0$. In this case the equation for L becomes a first order <u>ordinary</u> differential equation given by

$$\frac{dL}{dx} = - \frac{K_L}{u} L.$$

With the initial condition L(0) this readily yields the solution

$$L(x) = L(0)e^{-\frac{K_L}{u}x}. \tag{C.3}$$

Now substitute (C.3) into the equation for D derived in Exercise 6. This results in a linear non-homogeneous ordinary differential equation

$$u \frac{dD}{dx} - K_D(D_S - D) + K_L L(0)^{-\frac{K_L}{u}x} = 0.$$

If $S = D_S - D$, this becomes

$$- u \frac{dS}{dx} - K_D S + K_L L(0)e^{-\frac{K_L}{u}x} = 0.$$

Now employ the standard procedure of <u>variation of parameters</u> to obtain

$$S(x) = \frac{K_L}{K_D - K_L} \left(e^{-\frac{K_L x}{u}} - e^{-\frac{K_D x}{u}} \right) L(0) + S(0)e^{-\frac{K_D x}{u}}. \tag{C.4}$$

8. Compute the derivative of S(x) in the Streeter-Phelps equation in (A.1) and set it equal to zero. This leads to

$$x^* = \frac{u}{K_D - K_L} \ln \left\{ \frac{K_D}{K_L} \left(1 - \left(\frac{K_D - K_L}{K_L} \right) \frac{S(0)}{L(0)} \right) \right\}.$$

But, since $S(0) = 0$,

$$x^* = \frac{u}{K_D - K_L} \ln \frac{K_D}{K_L}$$

and substituting into (C.4) gives

$$S(x^*) = \frac{K_L L(0)}{K_D} e^{-\frac{K_L x^*}{u}} .$$

Therefore,

$$D(x^*) = D_S - S(x^*).$$

A straightforward computation shows that

$$\frac{d^2 S(x^*)}{dx^2} < 0$$

so that x^* provides a unique maximum for S (and hence, a unique minimum for D) assuming, as we do throughout, that $K_D - K_L > 0$. Because of uniqueness, the minimum of D on the interval $[0, x_0]$ is either $D(x^*)$ if $x^* < X_B$ or $D(x_B)$ when $x^* \geq x_B$.

9. Since $S(0) = 0$, the Streeter-Phelps relation above, together with the result of Exercise 8, shows that the minimum value of D on the interval from $x_A = 0$ to x_B is

$$D(x_B) = D_S - S(x_B) = D_S - \frac{K_L}{K_D - K_L}\left(e^{-\frac{K_L x_B}{u}} - e^{-\frac{K_D x_B}{u}}\right) L(0).$$

But $L(0) = L_A(1-\alpha)$. It follows that $D(x_B)$ is a linear function of α. Also

$$L(x_B) = L(0) e^{-\frac{K_L x_B}{u}}$$

as given by (A.1). But $L(0) = L_A(1-\alpha)$ and so again $L(x_B)$ is linear in α. Therefore, $\tilde{L}(x_B) = L(x_B) + L_B(1-\beta)$ is linear in α and β.

10. Since $D_{min} > \max(c_1, c_6)$ relations (A.2) give $\alpha \geq \text{constant}_1 > 0$, $c_7\alpha + c_8\beta \geq \text{constant}_2 > 0$. If both inequalities are strict, we can decrease α until either $\alpha = \text{constant}_1$ or until $c_7\alpha + c_8\beta = \text{constant}_2$ whichever occurs first. Since total cost is thereby lowered, it is clear that optimality requires that one of the inequalities in (A.2) be an equality.

11. $q_i = \dfrac{350\ u_{i1} + 3500\ u_{i2}}{10^6}$ mgd, $i = 3,4$.

12. From (B.3) we have

$$25{,}300\ z_3 = u_{31} + 10\ u_{32}.$$

Hence from (B.4),

$$350 \cdot 25{,}300\ z_3 = 10^6 q_3,$$

if q_3 is nonzero or, equivalently,

(C.5) $$q_3 = 8.85\ z_3, \qquad \text{if } z_3 > 5.$$

The smallest value q_3 can have is therefore $8.85 \cdot 5 = 44.25$ unless q_3 is zero (when $z_3 \leq 5$). Thus we impose the constraints

(C.6) $$\begin{cases} q_3 \geq 8.85\ z_3 - 44.25 \\ q_3(8.85\ z_3 - q_3) = 0 \end{cases}$$

together with the requirement that q_3 be as small as possible. If $z_3 \leq 5$, then the first relation in (C.6) allows $q_3 \geq 0$. (Recall: $q_3 = x_{31} + x_{34}$ and so is non-negative). Since q_3 is to be minimal, the second relation in (C.6) is satisfied only by $q_3 = 0$. However, if $z_3 > 5$, then the first relation of (C.6) forces q_3 to be strictly positive and so the second of (C.6) implies that $q_3 = 8.85\ z_3$. Conversely, if $q_3 = 0$, then it is apparent that $z_3 \leq 5$, while if $q_3 = 8.85\ z_3$, then $z_3 > 5$. Otherwise, (C.6) could be satisfied by the smaller value of $q_3 = 0$.

13. The revised equations are labeled as (15") through (22")

(15") $$\begin{cases} q_1 = x_{11} + x_{15} \\ \dfrac{350u_{31} + 3500u_{32}}{10^6} = x_{31} + x_{34} \\ \dfrac{350u_{41} + 3500u_{42}}{10^6} = x_{44} \\ q_5 = x_{51} + x_{55} \end{cases}$$

Relations (16), (17), (18), (19), (20), and (21) remain unchanged. Inequality (22) also remains except that D_S is given by

$$P_4 = \sum_{j=4}^{6} \beta_j t_{j4} + \frac{15u_{41} + 40u_{42}}{10^4}.$$

The cost term (23) needs to be augmented. Since we wish q_3 to be as small as possible, it suffices to add the term

$$u_{31} + 10u_{32}$$

to the cost D. One also needs to add on the constraints (B.1) and (B.2) as well as those developed in Exercise 12:

$$q_3 \geq 8.85\ z_3 - 44.25$$

$$q_3(8.85\ z_3 - q_3) = 0,$$

where q_3, z_3 are given by (B.3) and (B.4) in terms of u_{i1}, u_{i2}. Finally, one requires, in addition to $x_{ij} \geq 0$, that $u_{i1} \geq 0$ and $u_{i2} \geq 0$ for $i = 3,4$.

GPSR Compliance

The European Union's (EU) General Product Safety Regulation (GPSR) is a set of rules that requires consumer products to be safe and our obligations to ensure this.

If you have any concerns about our products, you can contact us on ProductSafety@springernature.com

In case Publisher is established outside the EU, the EU authorized representative is:

Springer Nature Customer Service Center GmbH
Europaplatz 3
69115 Heidelberg, Germany

Batch number: 10577876

Printed by Printforce, the Netherlands